FLORA OF TROPICAL EAST AFRICA

PONTEDERIACEAE

B. Verdcourt

Erect or floating aquatic herbs. Leaves with floating and/or emersed blades, sheathing at the base. Flowers hermaphrodite, regular or slightly irregularly two-lipped, arranged in racemes, spikes, panicles or sub-umbellate inflorescences subtended by a spathe-like leaf-sheath or solitary (*Eichhornia*). Perianth hypogynous, petaloid, divided into 6 lobes more or less arranged in two series, united at the base or free. Stamens 1, 3, or 6, equal or unequal, inserted on the perianth; filaments free; anthers dithecous, opening by longitudinal slits or rarely by terminal pores. Ovary superior, 3-locular with axile placentas or unilocular with 3 parietal placentas; style entire or lobed; ovules anatropous, numerous to solitary. Fruit a capsule opening by 3 valves, or an indehiscent achene. Seeds longitudinally ribbed, with copious endosperm.

A small family containing 7 or 8 genera, of which three occur native in East Africa and a further one is rather infrequently cultivated as an ornamental, namely *Pontederia lanceolata* Nutt. (Nairobi, *Verdcourt* 2418!, *Compton* H. 241/55!). *Eichhornia crassipes* (Mart.) Solms (Water Hyacinth) is a floating plant from South America which has become a great nuisance on many waterways throughout the tropics (but not where it is native) and is everywhere gazetted as a noxious weed. Since it is so important it has been included in the text, although it is not indigenous, nor well established save in one area.

Stamens 6; anthers dorsifixed:
 Perianth-segments free to the base; flowers distinctly
  pedicellate . . . . 1. **Monochoria**
 Perianth-segments united below into a tube; flowers
  sessile:
  Ovary 3-locular with numerous ovules; fruit a
   loculicidal capsule, seeds numerous; floating
   plants. . . . . 2. **Eichhornia**
  Ovary unilocular; fruit membranous, indehiscent
   with one pendulous seed; erect cultivated plant
   on margins of ponds; perianth blue with
   mustard-coloured blotch . . . . **Pontederia**
   (see note above)
Stamens 1–3; anthers basifixed . . . . 3. **Heteranthera**

1. MONOCHORIA

C. Presl, Rel. Haenk. 1: 127, 351 (1827); Solms in A.DC., Mon. Phan. 4: 522–525 (1883); Verdc. in Kirkia 1: 80 (1961)

Erect glabrous usually perennial aquatic herbs, sometimes with stems arising from a stout creeping rhizome. Leaves all radical with long petioles; flowering stems radical, each bearing one apical sheathing leaf, its petiole appearing as a prolongation of the stem. Flowers medium-sized, blue, pedicellate, arranged in racemes (sometimes subumbelliform); the inflores-

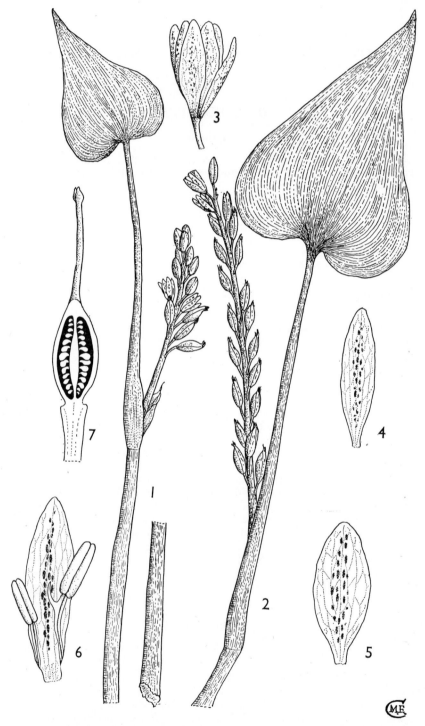

FIG. 1. *MONOCHORIA AFRICANA*—**1, 2,** flowering stems, × ⅔; **3,** flower, × 3; **4,** outer perianth-lobe, × 3; **5,** inner perianth-lobe, × 3; **6,** perianth-lobe with one each of the two types of stamen attached, × 4; **7,** gynoecium in longitudinal section, × 6. All from *Greenway & Rawlins* 9483.

cence is at first enclosed in a membranous spathe at the base of the peduncle, contained in the sheath of the apical leaf. Perianth-lobes 6, free almost to the extreme base, the inner 3 broader, at first spreading, later twisted. Stamens 6, fixed on the base of the perianth, equal or usually unequal, 5 with smaller yellow anthers, the sixth with a longer filament bearing an erect lateral tooth and having a longer blue anther. Ovary 3-locular; ovules numerous. Style filiform; stigma subentire or very shortly 3–6-fid. Capsule loculicidally splitting into 3 valves. Seeds oblong, longitudinally ribbed.

A small genus of about 5 species confined to the Old World, extending from tropical Africa to Manchuria and South Australia. Two species occur in tropical Africa.

M. africana (*Solms*) *N.E. Br.* in F.T.A. 8: 5 (1901); Schwartz in E.J. 61, Beibl. 139: 36 (1927); F.P.S. 3: 278 (1956); Verdc. in Kirkia 1: 81, t. 8 (1961). Type: Sudan Republic, Equatoria, Jur Ghattas, *Schweinfurth* 2296 (B, holo.†, K, iso.!)

Aquatic herb 0·5–0·9 m. tall; rhizome absent; roots fibrous, in a dense tuft. Leaf-blades ovate or elongate-ovate, acuminate, cordate or subcordate, 7·5–14 cm. long, 5–7·5 cm. broad; petioles up to ± 75 cm. long, narrowed above. Flowering stems ± 40–70 cm. long (including leaf); apical leaves similar to true radical leaves, probably more ovate (from the very limited herbarium material); petiole ± 17 cm. long. Inflorescence 5–14 cm. long, probably becoming longer in ripe fruiting state; peduncle 4–6 cm. long; spathe-like bract membranous, 1·6–2·5 cm. long, with a filiform point 2–8 mm. long. Flowers 15–30, prominently pedicellate; pedicels 3–7 mm. long. Perianth clear blue (*fide* Greenway) or violet (*fide* Schweinfurth), campanulate or somewhat funnel-shaped, ± 1·2 cm. long; outer lobes oblong, obtuse, 3 mm. broad; inner lobes elliptic, obtuse, 6 mm. broad. Stamens unequal, the longest with a tooth on the filament. Fruits not known. Fig. 1.

KENYA. Tana River District: S. bank of R. Tana, Ozi, 8 Nov. 1957, *Greenway & Rawlins* 9483!
DISTR. **K7**; Sudan Republic (Equatoria), South Africa (Transvaal)
HAB. Margins of rice-fields in cleared mangrove forest; 0–?1000 m.

SYN. *M. vaginalis* (Burm. f.) Kunth var. *africana* Solms in A.DC., Mon. Phan. 4: 525 (1883)

NOTE. N. E. Brown maintained that this plant is quite distinct from the Asiatic *M. vaginalis* (Burm. f.) Kunth, but Backer (Fl. Malesiana, ser. 1, 4: 258 (1951)) has again suggested that the African plant is a race of *M. vaginalis*. The rather striking differences in inflorescence lead me to believe that N. E. Brown was correct in separating the two, but very little material is to hand. In West Africa a further member of the genus occurs which Hutchinson & Dalziel (F.W.T.A. 2: 354 (1936)) referred to *M. vaginalis* (Burm. f.) Kunth var. *plantaginea* (Roxb.) Solms. The material I have seen is clearly different from *M. africana*, having much larger tepals and the inflorescences always overtopping the leaves, which are lanceolate without a cordate base; also the terete part of the petiole above the sheath is much shorter. I believe the W. African material represents a distinct species (*M. brevipetiolata* Verdc.).

2. EICHHORNIA*

Kunth, Eichhornia gen. nov. fam. Pontederiac. (1842) & Enum. Pl. 4: 129 (1843); Solms in A. DC., Mon. Phan. 4: 525–531 (1883), *nom. conserv.*

Piaropus Raf., Fl. Tellur. 2: 81 (1837)

Floating or creeping aquatic herbs, rooting at the nodes. Leaves often of two sorts, the floating or erect ones broad, lanceolate, ovate or orbicular, the submerged ones linear; petioles long, sheathing at the base or sheath pro-

* With Susan Carter.

duced at apex to form a " stipule ". Flowers solitary or in spikes or panicles arising from a convolute sheath-like spathe or bract, or fasciculate within the leaf-sheath. Perianth funnel-shaped, slightly irregularly divided into 6 lobes, the inner broader than the outer. Stamens 6, the 3 upper included, the 3 lower ± exserted, usually with longer filaments. Ovary 3-locular; ovules numerous; style filiform; stigma entire or very shortly 3–6-lobed. Capsule membranous, 3-locular, enclosed in the faded perianth, ovoid to linear-fusiform. Seeds numerous, oblong or ellipsoidal, finely ribbed.

A small genus of 7 species, one occurring in Africa, the rest natives of tropical or subtropical America, one of which, however, is now widely naturalized in the tropics of the Old World.

Leaves rosulate, erect; young petioles very swollen at
 or below the middle . . . . . . 1. *E. crassipes*
Leaves alternate, floating or submerged; petioles not
 swollen . . . . 2. *E. natans*

1. **E. crassipes** (*Mart.*) *Solms* in A.DC., Mon. Phan. 4: 527 (1883); Schwartz in E.J. 61, Beibl. 139: 34 (1927); Perrier, Cat. Pl. Madag.: 10 (1934); Täckholm & Drar, Fl. Egypt 2: 441–8 (1950); Backer in Fl. Malesiana, ser. 1, 4: 259, figs. 2 & 3 (1951); Jonker-Verhoef in Fl. Suriname 1: 85 (1953); Wild, Common Rhodesian Weeds, fig. 10 (1955), as "*Eichornea*", & in Kirkia 2: 9, frontispiece (1961); Ivens, E. Afr. Weeds: 3, fig. 3 (1967). Type: Brazil, Minas Geraes, *Martius* (M, holo.)

Floating aquatic herb ± 30–50 cm. tall with erect rosulate leaves and a dense tuft of plumose roots; reproducing vegetatively by means of stolons which give rise to new plants that readily separate; broken pieces of plant also form new plants rapidly. Leaves radical, erect, emerging from the water, rosulate; blades rhomboid or ovate, obtuse at the apex, cuneate to shallowly cordate, glabrous, very densely nerved, 5–25 cm. long and broad; petioles spongy, very much swollen at or below the middle, particularly in young plants, up to 30 cm. long; stipules 2–15 cm. long, with a blade-like portion above. Flowering stems with small apical leaf ± 2 cm. long and broad, its sheath large, ± 4 cm. long and 1·5 cm. wide, enclosing the spathe. Inflorescence with long peduncle, spicate, puberulous, of ± 3–35 flowers, 4–15 cm. long, bending over after fertilization. Perianth lilac or white, the upper lobe with a deep violet blotch containing a yellow spot; outer lobes narrowly elliptic; inner broadly elliptic, ± 3 cm. long; perianth-tube pubescent, 1·5–1·8 cm. long. Stamens unequal (see note); filaments pubescent. Capsule ± 1·5 cm. long. Seeds winged, 1·2 mm. long and 0·5 mm. wide.

KENYA. Nairobi, Aug. 1958, *Agric. Dept.* in *E.A.H.* 11830! (sterile) & Nairobi City Park, 5 May 1957, *Duncan* 1236!
TANGANYIKA. W. Usambara Mts, 6·5 km. NE. of Lushoto, Mkuzi, 11 June 1953, *Drummond & Hemsley* 2887!; Tanga District: R. Sigi near the Amboni Sisal Estate, from estuary for several kilometres upstream, Mar. 1958, *C. G. van Someren* in *E.A.H.* 11881! Also reported from R. Pangani, and Soni in W. Usambara Mts.
DISTR. K4, 6 (Narok); T3; native of Brazil, now widespread in the tropics of the Old World. Since declaration as a noxious weed in East Africa it is rarely seen. It is common in the Sudan Republic and will probably be found in Uganda.
HAB. Slowly flowing rivers, lakes formed by dams and small ponds; 0–1700 m.

SYN. *Pontederia crassipes* Mart., Nov. Gen. et Sp. 1: 9, t. 4 (1823)
  Piaropus crassipes (Mart.) Britton in Ann. N.Y. Acad. Sci. 7: 241 (1893)

NOTE. As *Eichhornia crassipes* is not a native plant, the full synonymy is not given here, but it may be found in the accounts of Backer and Jonker-Verhoef cited above; see also discussion and references given by Täckholm & Drar (1950). The plant is supposed to have trimorphic flowers but no observations have been made in East Africa. One form only occurs in Malesia with three very short anterior filaments and the others much longer. Wild (1961) discusses control measures.

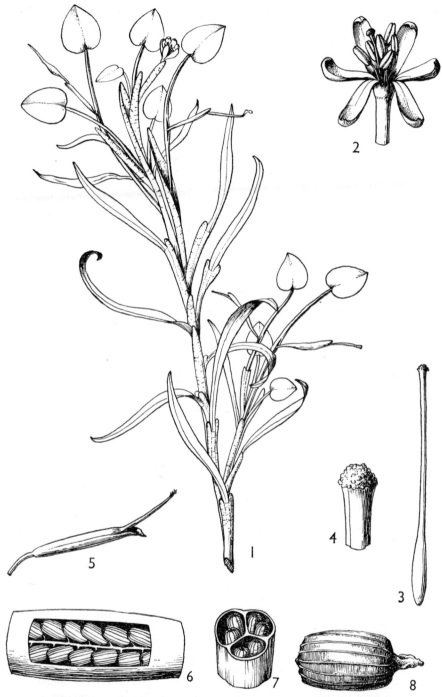

FIG. 2. *EICHHORNIA NATANS*—**1,** flowering stem, × 1; **2,** flower, × 4; **3,** gynoecium, × 4; **4,** tip of style and stigma, × 16; **5,** capsule and part of spathe, × 2; **6,** part of capsule, with wall partially removed to show seeds, × 8; **7,** transverse section of capsule, oblique view, × 8; **8,** seed, × 32. All from *Duke* 2.

2. **E. natans** (*P. Beauv.*) *Solms* in Abhandl. naturw. Ver. Bremen 7: 254 (1882) & in A.DC., Mon. Phan. 4: 526 (1883); N.E. Br. in F.T.A. 8: 4 (1901); Schwartz in E.J. 61, Beibl. 139: 33, 46, 50 (1927); Perrier, Cat. Pl. Madagascar: 10 (1934); F.W.T.A. 2: 354–5, fig. 303 (1936). Type: Nigeria, Oware (Warri) banks of R. Formosa, *Palisot de Beauvois* (G, holo.)

Floating aquatic herb with branched stems, sometimes rooting in the bottom soil. Leaves alternate, of two kinds: submerged leaves sessile, linear, acute, (2–)3–6·5(–10, *fide* F.T.A.) cm. long, (1–)2–3 mm. broad; floating leaves petiolate, ovate or orbicular, obtuse or rarely acute at the apex, cordate with small rounded overlapping lobes, minutely pustulate above, 1·3–3 cm. long, 1–2 cm. wide; true petioles 4–10 cm. long; stipules from 2 mm. in the submerged leaves to 3·2(–5) cm. long in the floating leaves, membranous, obtuse, sheathing the stem. Inflorescences 1-flowered, shortly pedunculate, arising from a tubular membranous obtuse spathe ± 1·3(–2) cm. long, apparently arising from below the middle of the petioles of the floating leaves but actually terminating short lateral 1-leaved branches of the main stem. Perianth somewhat irregular, blue or mauve (see note); lobes small, obovate, up to 4·5 mm. long, 2 mm. wide; tube 1·0–1·2 cm. long, elongating after anthesis to 2·2 cm. long. Stamens 6, unequal, glabrous. Style long and filiform. Capsule narrowly fusiform, 1·0–1·2 cm. long. Seeds numerous, oblong with many very fine longitudinal ribs, to 1 mm. long. Fig. 2, p. 5.

UGANDA. Acholi District: Madi, Dec. 1862, *Grant* 726! & 727! & Atiak, Nov. 1937, *Chorley* in *Chandler* 2055!; Teso District: Amuria, 14 Sept. 1946, *A. S. Thomas* 4540!
DISTR. U1, 3; W. Africa to the Congo and Sudan Republics, Zambia, Rhodesia, Angola, South West Africa, Madagascar
HAB. Stagnant pools in swamps, etc.; 1080–1300 m. (in Uganda)

SYN. *Pontederia natans* P. Beauv., Fl. Owar. 2: 18, t. 68/2 (1810); Bak. in Oliv. in Trans. Linn. Soc. 29: 162 (1875)
Monochoria natans (P. Beauv.) Thomson in Speke, Journ. Nile: 649 (1863)

NOTE. Until recently the African and New World plants have always been kept apart, either as distinct species, or at least as varieties (by Solms-Laubach). However, Jonker-Verhoef points out (Fl. Suriname 1: 85 (1953)) that although the African plants always have solitary flowers and the American plants nearly always have 2–3-flowered inflorescences, solitary flowers have been observed in some specimens from Suriname and on this basis unites them under *E. diversifolia* (Vahl) Urb.* He also states that the flowers are light violet-blue, the upper perianth-lobe bearing a yellow blotch; field notes of African plants indicate uniform blue colouration. Apart from these colour and inflorescence differences there is a marked difference in the size of the flowers. The American plant has the perianth 3 cm. long and 2 cm. in diameter whereas the African plant has the perianth under 1 cm. in diameter. After the examination of many specimens, both African and American, the present authors consider these and other differences to be sufficient grounds for maintaining the plants as distinct species.

3. HETERANTHERA

Ruiz & Pav., Fl. Peruv. et Chil. Prodr.: 9, t. 2 (1794); Solms in A.DC., Mon. Phan. 4: 516–522 (1883), *nom. conserv.*

Aquatic herbs; lower stems creeping and rooting in the mud. Leaves either all linear and submerged, or ovate or reniform and floating, sessile, or petioles long, sheathing the stem at the base; flowering stems bearing an apical leaf, the sheath of which encloses the membranous spathe. Flowers rather small, solitary or several, arranged in a spike subtended by the spathe; also one or more cleistogamous flowers enclosed in the spathe or sometimes mingled with the normal flowers. Perianth almost regular, funnel-shaped or salver-shaped, divided into 6 equal oblong spreading lobes. Stamens 3 in

* *Wright* 3266! (K) from Cuba is very similar to African material and probably represents the kind of specimen Jonker-Verhoef had seen.

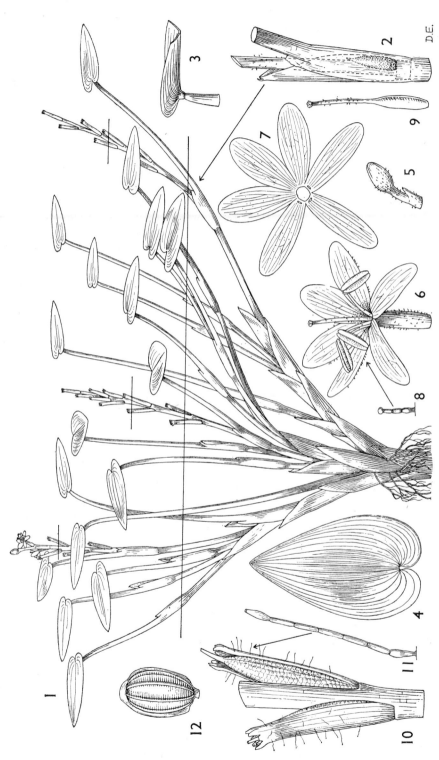

D.E.

FIG. 3. *HETERANTHERA CALLIFOLIA*—1, habit, × 1½; 2, detail of leaf-sheaths, × ½; 3, upper part of petiole and base of leaf-blade, × 1; 4, leaf-blade, surface view, × 1; 5, bud, × 4; 6, flower, × 4; 7, perianth, detached and flattened, × 4; 8, glandular hair from perianth, × 40; 9, gynoecium, × 4; 10, part of infructescence with two capsules, × 2; 11, glandular hair from perianth remnants sheathing capsule, × 40; 12, seed, × 22. 1-4, 10-12, from *Polhill & Paulo* 1704; 5-9, from *Polhill & Paulo* 2168.

normal flowers, sometimes solitary in cleistogamous flowers, inserted at the throat of the perianth-tube, exserted and unequal. Ovary unilocular with 3 parietal placentas or imperfectly 3-locular; ovules numerous; style filiform; stigma thickened. Capsule oblong or linear with thin pericarp. Seeds numerous, ovoid, ribbed.

A genus of about 10 species, two in Africa the rest in tropical and subtropical America. One species is sometimes cultivated in Asia. Only one occurs in the Flora area.

H. callifolia *Kunth*, Enum. Pl. 4: 121 (sphalm. 123) (1843), as " *callaefolia* "; Solms in A. DC., Mon. Phan. 4: 521 (1883) as " *callaefolia* "; P.O.A. C: 137 (1895); N.E. Br. in F.T.A. 8: 2 (1901), as " *callaefolia* "; F.W.T.A. 2: 355 (1936); Robyns & Tournay, F.P.N.A. 3: 339 (1955); F.P.S. 3: 276 (1956); Verdc. in K.B. 1957: 352 (1957). Type: Senegal, *Sieber* 51 (P, holo., K, iso.!)

Glabrous aquatic plant ± 13–50 cm. tall; lower stems creeping and rooting in the mud, sometimes almost free-floating; stoloniferous. Leaves erect or ascending or sometimes floating, broadly ovate, subobtuse or shortly and obtusely pointed, cordate, 2·5–7·5 cm. long, 1·3–5 cm. wide; petioles 5–20 cm. long. Flowering stems ± 3·8–10 cm. long, with an apical leaf similar to the ordinary leaves. Inflorescence a spike 5–10 cm. long of ± 7–20 small sessile flowers, with a membranous spathe at its base subtended by the leaf-sheath; this spathe contains a basal cleistogamous flower. Perianth white, blue or purple; lobes spreading, oblong, obtuse, ± 4–5 mm. long; tube cylindrical, 5–10 mm. long. Stamens 3, shortly exserted, sometimes only one in cleistogamous flowers. Capsule oblong, trigonous, 0·8–1·2 cm. long or (in the cleistogamous flower) 1·2–1·8 cm. long. Fig. 3, p. 7.

UGANDA. Acholi District: Madi, Dec. 1862, *Grant* 655!; Mengo District: Bugerere, 8·5 km. N. of Bale, 3 July 1956, *Langdale-Brown* 2147! & Kampala, Ngulu Hill, Aug. 1937, *Hancock & Chandler* 1842!
KENYA. Kisumu-Londiani District: near Kisumu, 9 June 1958, *McMahon* in *E.A.H.* 11831! & Ahero, 11 Aug. 1958, *McMahon* 37!
TANGANYIKA. Mbulu District: Lake Manyara National Park, 26 Feb. 1964, *Greenway & Kanuri* 11257!; Dodoma District: 9·6 km. from Manyoni on Itigi road, Aghondi, 25 Apr. 1962, *Polhill & Paulo* 2168!; Iringa District: 6·4 km. N. of Iringa, 10 Mar. 1962, *Polhill & Paulo* 1704!
DISTR. U1, 2, 4; K5; T1, 2, 4, 5, 7; W. Africa, Congo and Sudan Republics, also southwards to N. Transvaal and South West Africa
HAB. Swamps and pools; 1000–2575 m.

SYN. *H. kotschyana* Solms in Schweinf., Beitr. Fl. Aethiop.: 205, 294 (1867) & in A.DC., Mon. Phan. 4: 522 (1883); N.E. Br. in F.T.A. 8: 3 (1901). Type: Sudan Republic, Kordofan, by Mulbes at Obeid, *Cienkowsky* 378 (W, holo.)
[*Monochoria vaginalis* sensu auctt., *non* (Burm. f.) Kunth]

INDEX TO PONTEDERIACEAE